Pygmy Goats as Pets.

Pygmy Goats, Mini Goats or Dwarf Goats: facts and information.

Raising, breeding, keeping, food, care, health and where to buy all included.

by

Elliott Lang

Published by IMB Publishing 2013

Acknowledgements

I would like to thank my children for inspiring me to write this
book. Their love and devotion for our
pygmy goats, AJ and Buckley, made me want to share the
joy of keeping these pets with the world.

Additional thanks to my wife, whose patience
with me knows no bounds.

Table of Contents

Chapter One: Introduction

When most people hear the word "pet" they picture a cuddly kitten or a fluffy dog – maybe even a hamster or a bird. Most people do not associate the image of a horned and bearded goat with the word "pet". Pygmy goats are, however, becoming very popular as pets and for good reason. These animals are very friendly and can be incredibly entertaining. If you have ever wanted a pet but don't want to go for something "traditional" like a cat or dog, a pygmy goat might be the right choice for you.

While these animals may not be the first thing that comes to mind when you think of the word "pet," they actually make great companion animals. Pygmy goats are incredibly intelligent and affectionate by nature, which makes them a joy to keep. Additionally, the cost and care required for keeping these pets is significantly lower than keeping other kinds of livestock. If you are looking for an entertaining and unique pet, consider the pygmy goat.

In this book you will learn the basics about pygmy goats including their origins as pets, what they look like and how to care for them. You will receive all the information you need to prepare your pygmy goat enclosure and to stock up on necessary supplies to keep your goats healthy. You will also find a number of valuable tools and resources regarding feeding, breeding and general care of pygmy goats as well as important information regarding preventing and treating pygmy goat diseases.

As an added bonus, you will also receive a wealth of information about showing pygmy goats in addition to a long list of frequently asked questions. Everything you have ever wondered about pygmy goats can be found within the pages of this book! If you've ever considered keeping pygmy goats as pets, this book is just the thing you need to get started.

What are you waiting for?
Start learning about pygmy goats today!

Chapter Two: Understanding Pygmy Goats

1.) What Are Pygmy Goats?

The pygmy goat is a breed of miniature domestic goat that is often kept as a meat goat. These goats also do well as milk producers and they are sometimes kept as pets. Pygmy goats are hardy creatures and are easily adaptable to a variety of climates and settings. Though similar in appearance to other breeds of domestic goats, the pygmy goat exhibits a few anatomical differences including the existence of dewclaws as well as a thurl, the type of hip joint commonly found in cattle.

2.) Facts About Pygmy Goats

Pygmy goats typically weigh between 50 and 90 lbs. Females of the species, called does, generally weigh between 53 and 75 lbs. (24 to 34 kg) while males, called bucks, weigh between 60 and 90 lbs. (27 to 39 kg). Most pygmy goats reach a maximum height at the withers between 16 and 23 inches (41 to 58 cm). Pygmy goats display a wide variety of colors including white, black, grey, brown and several variations on caramel and grey agouti.

Pygmy goats are hardy animals and they tend to be very good-natured and docile. These features are what make the pygmy goat a popular pet. In addition to its friendly nature, the pygmy goat is also a good provider of milk and an efficient browser. These goats do well in a variety of settings and they are capable of adapting to almost any climate.

These goats do not require a great deal of special care. In fact, they will do just fine provided with little more than an 8x10-foot shed furnished with sleeping and feeding areas. Given adequate space to roam and graze, these goats make very active and sociable pets. Not only are these goats fairly simple to keep, but they are also precocious breeders – females generally bear one to four kids every 9 to 12 months.

3.) History of Pygmy Goats as Pets

Pygmy goats originated in West Africa in the Cameroon Valley. There are two types of dwarfism evident in goats in Africa. Achondroplasia results in disproportionately short legs with a plumb body and a short head while pituitary hypoplasia results in a small goat with normal proportions. The first of these two types is more common in West African goats while the second type is common throughout the southern Sudan.

In Britain, the Pygmy Goat Club does not differentiate between the two types, though it is thought that modern pygmy goats are more closely related to the dwarf goats of West African than to those from the southern Sudan. Prior to the formation of the club in 1982, dwarf goats were identified by regional names such as Cameroonian, Nilotic, Nigerian, West African and Sudanese. The Pygmy Goat Club in Britain discarded these regional names, however, in favor of the generalized name Pygmy Goat.

It is thought that these goats were domesticated as early as 7,000 B.C. but they weren't imported from Africa into European zoos until the 1950s. Pygmy goats were kept on display as exotic animals in the zoos of Sweden and Germany and were sometimes used as research animals. In 1959, the first shipments of pygmy goats were sent from Sweden to the United States. The recipients of these shipments were the Rhue family in California and the Catskill Game Farm in New York State. Over time, the popularity of this species spread and they were soon kept by private breeders as pets and exhibition animals. Today, the pygmy goat is a common sight in petting zoos and they are also widely kept as pets in both North America and Europe.

a.) National Pygmy Goat Association

The National Pygmy Goat Association (NPGA) is the official breed association of the pygmy goat in the United States. This association was formed in 1975 with the goal of "supporting the Pygmy Goat in the United States by collecting and disseminating information, protecting the breed standard, and recording their lineage through [their] registry." As part of their mission, the NPGA is also dedicated to facilitating communication between researchers and breeders in addition to establishing regional affiliate clubs. The NPGA also strives to promote and popularize the breed through publications.

Members of the NPGA must pledge to uphold a code of ethics, citing their commitment to upholding the breed standard and to abide by show rules. NPGA members must also pledge to be truthful and accurate in advertising and to only sell animals that are in good condition. Members of the NPGA have access to a wealth of information on the association's website and may also search or post listings for pygmy goats for sale. Other information found on the site includes show rules and schedules, an online herd book, NPGA merchandise and valuable resources for information regarding the care and keeping of pygmy goats as pets and for show.

b.) Origins of the Pygmy Goat Club

The Pygmy Goat Club is the official pygmy goat association of Great Britain. This group was started by breeder Sylvia Collyer from Alton, Hampshire. Collyer maintained a herd of blue-roan pygmy goats and, in 1981, wrote a letter to *Fur and Feather* magazine inquiring whether any other pygmy goat breeders were interested in starting a club. A year later, the club was officially launched and an Identity Register for pygmy goats was opened. The goal of this register was to compile information regarding pedigree details of Pygmy goats in the country.

The early meetings of the Pygmy Goat Club were dedicated to establishing a breed standard and to establishing the pygmy goat as a breed rather than classifying it by different regional names. The breed standard was established based on the American breed standard, allowing all colors and markings with the exception of pure white and Swiss markings. After establishing the breed standard, Collyer moved on to writing a Pygmy Goat Booklet. The first Pygmy Goat Club show was held at Chelmsford Cattle Market in May of 1985 and it was judged by George Starbuck. The Pygmy Goat Club continues to change and grow and it is widely regarded as an established breed society.

4.) Types of Pygmy Goats

Pygmy goats are known by the scientific name *Capra hircus*. These goats exhibit a wide range of colors and patterns, but all pygmy goats belong to the same species. Pygmy goats have a thick body with a heavy coat supported by thick, sturdy legs. The snout is long and the long ears fold forward over the head.

Genetics play a significant role in determining the coloration of pygmy goats and all colors are acceptable.

The coat of pygmy goats is typically full and straight, made of medium-long hair – the density of the coat may vary from one season to the next. Adult males exhibit abundant hair with a long, full beard. Males may also exhibit a copious mane draping from the shoulders. Female pygmy goats may exhibit a beard, but it is often sparse or trimmed. The most common coloration for pygmy goats is an agouti (grizzled) pattern composed of an intermingling of dark and light hairs.

Though pygmy goats of any color are acceptable, they must exhibit breed-specific markings in order to meet conformation standards. Solid black goats are acceptable, but multi-colored goats must exhibit coloration on the muzzle, forehead, ears and eyes that is lighter than the main portion of the body. The crown, dorsal stripe and martingale are darker than the main body color as well as the front and rear hooves. Goats exhibiting caramel coloration typically exhibit light-colored vertical striping on the front of dark-colored socks.

Summary of Pygmy Goat Facts

Average Lifespan: 10 to 15 years
Average Weight (Female): 53 to 75 lbs. (24 to 34 kg)
Average Weight (Male): 60 to 90 lbs. (27 to 39 kg)
Average Height: 16 to 23 inches (41 to 58 cm)
Common Coat Pattern: agouti (grizzled)
Acceptable Coat Colors: any
Breed-Specific Markings: lighter muzzle, forehead, ears and eyes; darker crown, dorsal stripe, martingale, hooves

Chapter Three: What to Know Before You Buy

Before you go out and purchase a pygmy goat, it is important that you understand a few things. First, you may be required to have a license or permit to keep pygmy goats on your property – failure to obtain this permit could result in fines or legal action. You should also take the time to determine whether pygmy goats will get along with other animals you have on your property and how many of them you should keep.

In addition to this basic information, it would also be wise to familiarize yourself with the costs associated with keeping pygmy goats so you can determine whether it is a practical venture for you or not. You will find all of this valuable information in this chapter.

1.) Do You Need a License?

In the United States, a permit is generally required to keep pygmy goats as pets. In order to receive a permit, you will have to fill out an application and file it with the Office of Animal Control. You will be required to include in the application the breed and gender of the goat(s) as well as a photo of what the animal looks like. You may also need to include proof of vaccinations and health examinations. If you intend to keep pygmy goats on rented property, you will also need written permission from the property owner.

Some states may also require you to post a sign for a certain period of time informing your neighbors that you will be keeping goats. If any written objections are received by the Office of Animal Control, a public hearing may be held. Keep in mind that a permit for pygmy goats only allows for a certain number of goats – generally no more than two plus offspring less than 6 months of age on lots up to 20,000 square feet.

In addition to limiting the number of goats you can keep on your property, a permit may also require you to construct a fenced-in space to keep the goats. This area must be properly fenced, drained and clean at all times. Some areas may allow pygmy goats to be kept in unenclosed areas as long as they are bound by a leash or tether so they cannot exit the property. The permit may also require that your pygmy goats be dehorned and neutered. Permits generally need to be renewed each year. Contact your local council to receive more specific information regarding permits in your area.

In the United Kingdom, permit requirements may vary according to the number of pygmy goats you intend to keep and the purpose for which you are keeping them. You may need to submit a holding register with the Department for Environmental Food and Rural Affairs (defra) in addition to electronically tagging all of the pygmy goats you are keeping. You may be required to provide specific information regarding the location of the holding as well as the species and purpose of keeping the pygmy goats.

In order to receive a permit, you might need to register as a keeper of livestock and receive a flock/herd number that will be used to identify your goats. Not only will you need a permit to keep your pygmy goats, but a permit may also be required if you intend to move them (such as for show). Contact your local council or defra directly for more specific information regarding the permits needed to keep pygmy goats in the UK.

2.) How Many Should You Buy?

Pygmy goats are very social creatures by nature so it is recommended that you keep them in groups. The number of pygmy goats you should buy will be determined by how much space you have. For each pygmy goat you intend to keep, you should be prepared to offer 15 to 20 square feet – that is an area of about 4 by 5 feet. You should also make sure you have enough indoor space for all of the goats you intend to keep.

Keep in mind that you will need to keep does and bucks separated, so this may be a factor in how much space you have to accommodate your pygmy goats. If you plan to keep multiple male goats, you will need a barn with separate stalls and outdoor pens to keep the bucks separated. To provide the bucks with company, you might consider keeping a buck kid in the stall with the adult buck – this will provide the adult with company and the kid with comfort and security.

If you are able to keep your pygmy goats in open pasture, this is the ideal situation. When this isn't possible, do your best to provide your goats with ample spacing in fenced pens. A 30-by-30 foot pen is adequate for two to four goats but more space is always better than less. Even if you do keep your goats in a pasture, you still need to provide ample shelter space for each goat you keep.

3.) Can Pygmy Goats be kept with Other Pets?

Pygmy goats are very friendly animals and they generally do well as companion animals for other species of livestock. For example, pygmy goats are often seen at racetracks and in horse breeding facilities. They have also been used in zoos as companion animals for elephants. Pygmy goats may also get along with household pets including dogs and cats.

Not only can pygmy goats be companions for other animals, but they also make good companion pets for people. These goats are gentle and affectionate – their small size also makes them suitable for companionship with children. Pygmy goats have also been known to serve as therapy animals.

4.) Cost of Care

Keeping pygmy goats is generally less expensive than keeping other types of livestock, but it all depends on how many you keep and the quality of the materials you choose for housing, fencing and feeding. If you choose very basic materials for housing and fencing, you can greatly reduce the cost of keeping pygmy goats but you may need to repair or replace the items sooner than you would if you invested in high-quality materials. Feeding pygmy goats is fairly inexpensive but veterinary costs for vaccinations, deworming and exams may vary considerably.

When you first decide to raise pygmy goats, you should expect to pay some initial costs. These costs will include the price of the goats themselves as well as the cost of building the enclosure, providing shelter and stocking up on food. After these initial costs are covered you will then need to provide your goats with food and veterinary care on a regular basis. The following sections will help you to estimate the cost of raising one or more pygmy goats.

a.) Summary of Initial Costs

Purchase price for pygmy goats is generally between $150 and $350 each (between £90 and £215) . The price may vary, however, according to whether you are purchasing from a registered breeder or from an online ad. Price may also vary depending whether you are purchasing a male or female kid. Costs for building an enclosure and shelter for your pygmy goats can be extremely variable depending on how many goats you intend to keep and what kind of space you are working with. Enclosures may cost anywhere from $200 to $1000 depending on the quality of materials and size.

In addition to purchasing the pygmy goats, you may also need to pay for certain veterinary services at the outset. Before you bring your pygmy goats home you need to make sure they have been properly dehorned, vaccinated and dewormed. These services can cost anywhere from $30 to over $100 (from £18 to £ 62) depending whether you get them from a veterinarian or directly from the farm. You may also want to consider additional costs such as membership to your regional or national pygmy goat club or association in addition to any fees required for attaining or renewing a license to keep pygmy goats on your property.

b.) Summary of Monthly Costs

After covering the initial costs with regards to purchasing and preparing an enclosure for your pygmy goats, you will also need to factor in the monthly costs for feeding your goats as well as for veterinary care. Standard veterinary exams may cost anywhere from $20 to $40 (from £12 to £24) per visit, not including the cost of medications and vaccines which can cost as little as $2 each or as much as $30 (£18) per month. Feeding costs will vary depending on the number of pygmy goats you keep. A bale of alfalfa hay may be priced between $9 and $12 (between £5 and £8) per bale and grain generally runs at about $10 (£6) per 50-pound bag. Other monthly costs to consider include utilities such as heating and water for your pygmy goat shelter – these costs will vary depending on the number of pygmy goats you have and the size of your shelter.

5.) Pros and Cons of Pygmy Goats

Pros of Pygmy Goats

a) The diminutive size of pygmy goats means they don't require as much space as other livestock
b) Very friendly and affectionate by nature
c) Make great pets, especially for children and families
d) Serve well as companion animals and even service animals
e) Easy to train – can be used for milking or cart-pulling
f) Breed all year round – kids are available any time of year
g) Fairly easy to breed – it does not require any special training to breed pygmy goats

Cons of Pygmy Goats

a) Pet goats of any kind can be messy and destructive
b) Love to jump and climb, may escape from enclosure if fences are too low
c) May be picky eaters – requires a certain diet
d) Unplanned pregnancies may be an issue if you fail to separate kids at the proper age
e) License is required to keep, may come with restrictions
f) Intact male pygmy goats can be very temperamental and may have a pronounced odor
g) Cost of vaccinations and treatment of disease can become expensive
h) Male and female goats should be kept separately – may require the construction of additional enclosures

Chapter Four: Purchasing Pygmy Goats

1.) Where to Buy Pygmy Goats (US and UK)

When it comes to buying pygmy goats, you have a variety of options to choose from. If you are not concerned about buying kids, your starting point should be to check local goat rescue centers in your area. Goat rescues in the U.S. and UK take in abandoned goats and rehabilitate them before offering them to new owners. You can find goat rescues in your area by searching online or by contacting your local county animal services.

If you prefer to purchase from a registered breeder and you live in the U.S., you may want to check the National Pygmy Goat Association (NPGA) website. In addition to offering valuable information about the care and show of pygmy goats, the NPGA also gives registered breeders an opportunity to advertise the sale of stock on their website. You may also be able to find local breeders in your area using the website.

In the UK, you can find a list of registered breeders at the Pygmy Goat Club website. Though the breeders on the list are members of the Pygmy Goat Club, the goats themselves may not be registered so it is important that you clarify this with the individual breeder. You can also contact your local county animal services for information regarding pygmy goat rescues in your area. You may also be able to find listings for pygmy goats by individual owners online.

National Pygmy Goat Association - U.S. Breeders:
http://www.npga-pygmy.com/contacts/breeders

Pygmy Goat Club – UK Breeders:
http://www.pygmygoatclub.org/breederslist.htm

2.) How to Buy Pygmy Goats

When purchasing pygmy goats, it is important to determine that you are buying from a responsible breeder and that the goats you are purchasing are healthy. You should also take the time to learn whether the goats have been vaccinated, what they are currently being fed and the age of the animal.

The following questions are important to ask when purchasing a pygmy goat:

How old is the Goat?

Many pygmy goat owners prefer to buy goats when they are kids – this ensures that you are able to raise the goat yourself. Before buying, ask the seller how old the kids are. Kids should not be weaned before 8 weeks of age – many sellers will not sell kids until they are 12 weeks of age. If you purchase a kid that is 8 weeks old or younger, you may need to bottle feed the kid – this can be very difficult if the kid is used to feeding directly from its dam.

What is the Goat Eating?

Even though pygmy goat kids may not be fully weaned until 10 to 12 weeks of age, they should still be eating grain and hay before that point. It is important to determine that the kids you buy are used to eating these foods before you bring them home. It is also wise to ask exactly what type of grain and hay the kids are eating because sudden changes in diet can cause stomach upset in goats. It is best to feed the kids the diet they are used to when you first bring them home and then slowly change the diet, if necessary, over the course of 7 to 10 days.

Has the Goat Received Supplements?

Some pygmy goat breeders choose to offer their goats vitamin and mineral supplements to improve their growth. While supplementation is generally not necessary, it can be beneficial if certain nutritional deficiencies are common in your area. Ask the breeder about what deficiencies your pygmy goats may be at risk of and determine whether they have been receiving supplements or not. If the goats have been receiving supplements, you may want to continue offering them or slowly wean the goats off the supplements to prevent problems.

Has the Goat been vaccinated?

Vaccinating pygmy goats is incredibly important, especially if you are adding new goats to an existing herd. If the goats have not been properly vaccinated, they could spread disease to the goats you already have. Before you buy, ask about the vaccination and deworming history of the goat and ask for a written copy of the history as well. Pygmy goats should generally receive their first

vaccine for clostridial diseases (CDT) by 10 to 12 weeks of age –
they should be started on a parasite control program by this time.

Has the Herd Been Tested for Disease?

Even if you are able to determine that the goats you are
purchasing have been vaccinated properly, it is still wise to ask
about the health of the herd. Diseases affecting pygmy goats can
take weeks or years to show up and, by that time, they are difficult
if not impossible to cure. Some diseases are highly contagious and
some can even be transmitted to humans. Ask the seller whether
the herd is tested regularly for disease so you can be sure you are
buying a healthy goat.

Is the Goat Registered?

When purchasing goats as kids, you may not be able to tell
whether the goat you are buying is truly a pygmy goat or not. If it
isn't, you may find yourself with a 200-pound dairy goat on your
hands instead of the pygmy goat you were expecting. To prevent
this from happening to you, make sure to purchase from a
registered breeder.

Chapter Five: Caring for Pygmy Goats

Keeping pygmy goats as pets can be a very enjoyable experience but it is important that you understand the basics about caring for them. Like all creatures, pygmy goats require food, water and shelter and it is your responsibility to provide for those basic needs. In this chapter you will learn how to provide adequate shelter, care and food for your pygmy goats – you will also learn the basics about breeding pygmy goats.

1.) Care and Requirements

Pygmy goats are known for being very adaptable to a variety of climates and environments. For this reason, pygmy goats are relatively easy to care for as pets. Like all animals, however, pygmy goats do have certain requirements in terms of housing, feeding and general care. When constructing an enclosure for pygmy goats, there are a number of things you need to consider.

The first consideration is space – for each goat you plan to keep, you will need at least 15 to 20 square feet of open space. This is a minimum estimate simply for the comfort of the animal – be sure to check your local permit/licensing requirements to determine how much space you legally need to provide per goat. You should also plan to enclose the area in fencing of at least 4-feet high. Though pygmy goats may look small, they are capable of jumping to surprising heights.

Shelter is also a key concern in constructing an enclosure for pygmy goats. In Southern states where the climate is consistently warm, you may need a little more than a basic shed-type structure. In Northern areas, however, you may need a barn to protect your goats from cold, wind, rain and snow. In constructing a barn, be sure to factor in considerations like size, accessibility and maintenance. The barn should be large enough to comfortably accommodate all of your goats and it should be easily accessible to you and to them. You should also consider how easy it will be to clean the barn.

In addition to providing basic shelter, pygmy goats also enjoy having niche areas to sleep and play. Feel free to include raised structures and platforms built into the walls. Goats love to jump up onto these structures to sleep and it will also provide a place for young kids to play. You should also be sure to include access to exercise yards that connect directly to indoor stalls so your goats can come and go as they please.

The type of flooring you choose for your barn is very important. If you use traditional wood flooring, it is likely to hold the odor of urine and will eventually rot. Cement floors are easy to maintain but they also can be cold and damp. Perhaps the best type of flooring for pygmy goat barns is a gravel base covered in a thick layer of clay. The clay will become compacted by the goats' hooves making it easy to sweep out and any moisture will seep through the clay into the gravel layer.

Additional considerations in caring for pygmy goats include providing plenty of fresh water. It is very important to keep your goats hydrated but realize that if the water is not fresh your goats are unlikely to drink it. Don't give in to the temptation to save money by purchasing cheap fencing or building materials. Remember that the fencing and shelter you provide for your goats does more than keep them in – it also serves to keep dangerous predators out.

Summary of Care and Requirements

Space: at least 15 to 20 square feet per goat (check your local permit/licensing requirements)
Shelter: enclosed barn in Northern climates, shed in Southern climates
Flooring: gravel base covered in thick clay
Extras: raised platforms for feeding and sleeping

2.) Breeding Pygmy Goats

Breeding pygmy goats is fairly easy because these creatures are naturally precocious breeders. Under the right conditions, female pygmy goats will produce between 1 and 4 kids every 9 to 12 months following a 5-month gestation period. Female goats intended for breeding are referred to as nannies and they are generally bred for the first time between 12 and 18 months of age. It is possible, however, for nannies to conceive as early as two months so it is important to keep nannies separate from male kids early on.

At birth, kids generally weigh between 2 and 4 pounds and they are able to jump and run within four hours of being born. Newborn kids begin nursing immediately and will begin eating roughage and grain within one week. Kids are generally weaned at 12 weeks (3 months) and are considered mature after 8 to 12 months. Pygmy goats exhibit polyestrous sexual behavior, which means that females experience heat and they can be made to come into milk production year-round. Goats that are kept primarily for milking can be made to provide a continuous supply of milk by alternately breeding two pygmy does.

a.) Guide for Breeding Pygmy Goats

Breeding pygmy goats can be a very rewarding experience and it is generally not difficult to accomplish. Pygmy goats are year-round breeders so they are capable of producing kids twice per year. As long as you wait to breed your goats until they have reached the proper age, you should have no problem encouraging your female goats to accept a male.

Keep in mind that it is unwise to house male and female pygmy goats together after the age of 2 months unless the males have been castrated. If you do intend to keep one or more intact males, it is best to keep them completely separate from the female goats. Unless you intend to breed your goats regularly, you might want to look into using a stud male in your area rather than keeping an intact male yourself.

It is best to wait until your female goats have reached an age of 15 to 18 months prior to breeding. Prior to breeding the does, it is wise to drench them in an anthelmintic (de-worming solution) and to do so again a few days after kidding. During the first 3 ½ months of pregnancy, does should be offered the same diet as usual. If you do not feed your goats individually but prefer a communal feeding style, make sure your pregnant does get their fair share of the food

After the 3 ½ month mark, you may want to consider separating your pregnant does into individual pens. You may also need to slowly increase their rations of concentrated feed to about double the usual amount. One month before kidding, vaccinate the does for tetanus and enterotoxemia – this will provide protection for the kids until they are old enough to receive the vaccines themselves. As your pregnant doe's due date approaches, clean the pen and raise the water bucket off the floor so newborn kids do not fall in and drown.

About 10 days before your doe's due date, begin checking her regularly for signs of labor. Look for signs that your goat's udder is filling with milk and that the teats look shiny. The muscles on either side of the spine at the doe's rump may also slacken and the goat may begin to paw at the bedding in her pen. It is also important to maintain a healthy diet during this stage because your goat will need energy during the birthing process.

When your goat is ready to give birth, the cervix will dilate and it will expel a thick mucus plug. As your goat begins to experience contractions, her body will stiffen and she may stretch out or dip her back. As the kid enters the birth canal, the doe will begin to strain during the contractions – some goats stand during this process but many prefer to lie down. As the doe continues to strain, you will see a water bag (amniotic sac) emerge from the birth canal containing the baby goat.

Once the kid has been born, you may need to break the water sac if it does not break on its own. You should then check the nose and mouth of the newly born kid to be sure it is clear of mucus so the kid can breathe. After severing the umbilical cord, spray it to prevent infection and give the doe some time to bond with the kid. Once she has, clean the kid with old towels and rub it dry.

After the kid has been born and cleaned, refresh the bedding in your goat's pen and provide fresh water. Make sure the kid has found its way to the doe's teat and has begun to suckle. Be aware that it may take several hours after birthing for the placenta to pass – do not rush this process and remove the placenta as soon as it has been passed.

b) Milk Replacer for Pygmy Goats

In the event that you do not have a lactating female pygmy goat, you may need to make some milk replacer. This formula can be used to supplement the diet of pygmy goat kids or it can act as a replacement for milk if the kid has been orphaned. When using milk replacer, only feed the kids about 1 oz. for each ¼ lbs. of body weight. Divide the feeding into three or four feedings and increase the feeding amount as the kid grows.

Milk Replacer Recipe:

2 cups pasteurized goat's milk
½ cup heavy cream

**Combine the ingredients in a bottle and shake well to combine. You may choose to warm the milk slightly to make it more palatable to your pygmy goat kids.

c.) Keeping Male Goats

If you do not plan to breed your pygmy goats on a regular basis, you should seriously consider whether or not you truly want to keep male goats. Intact male goats are known for having a penetrating odor and for being very temperamental. If you are not prepared for these things, you may have problems with your goats. If you do plan to keep male goats, consider having them castrated to avoid these problems. In the event that you do decide to breed your goats, ask around for local stud males available for breeding in your area. This is a much better alternative to keeping a full stud male yourself, unless you have a very active breeding program in place.

d.) Pygmy Goat Breeding Summary

Gestation Period: 145 to 155 days (approx. 5 months)
Heat (Oestrus) Cycle: 18 to 24 days
Length of Heat: 12 to 48 hours
Weaning Age: 8 to 10 weeks
Sexual Maturity (Male): 10 to 12 weeks
Onset of Heat (Female): 3 to 12 months

3.) Feeding Pygmy Goats

Pygmy goats are ruminant animals, which means that their stomachs have four compartments: the rumen, reticulum, omasum and abomasum. The rumen is the compartment that contains micro-organisms that work to ferment the food the goat eats. While eating roughage such as hay or other plants, goats add saliva to the roughage and swallow it. Later, the material is regurgitated and chewed a second time – this process is called rumination.

Newborn goats primarily utilize the abomasum compartment of the stomach – the part that corresponds to the human stomach. Kids immediately begin to nurse after birth and the milk consumed goes directly into the abomasum. After a week or so, kids generally begin to eat leafy hay and the rumen compartment of the stomach begins to develop. It can take up to 8 to 10 weeks, however, for the rumen to become fully developed. Before the rumen is fully developed, baby goats function like animals that have only a single stomach.

Pygmy goats are grazing animals – they tend to forage for food, consuming various plants, grasses and brushes. Goats generally prefer broad-leafed plants and weeds including dandelions and clovers, but they will also eat wild grasses. When it comes to hay, goats tend to prefer legume hay (such as alfalfa or clover hay) to grass hay.

During the summer months, pygmy goats being kept as pets and not for heavy milking generally do not require food other than pasture grazing. Does used for milk production and young goats will, however, need supplementary food for energy. Grains including oats, corn and barley are good energy sources for goats and they do not need to be ground before feeding. Most goats prefer rolled grains to ground grains and rolled grains are generally recommended for young goats so they learn to chew grain properly at an earlier age.

Clean, fresh water is also a necessity for keeping pygmy goats. In addition to fresh water, pygmy goats also require certain nutrients in order for their bodies to be able to perform necessary functions. Some of the most important nutrients for pygmy goats include protein, carbohydrates, fat, vitamins and minerals. The protein needs of pygmy goats can be met through alfalfa and clover hay or through soybean meal. It is important to note that young goats and pregnant does require more protein than other goats.

Carbohydrates and fats provide pygmy goats with energy. These nutrients generally come from whole or rolled grains like those mentioned above. You should be wary in feeding your goats mill by-products because they may be lower in fat and carbohydrate than whole grains. As is true of protein, young goats and lactating or pregnant does have higher energy needs, which are best met with whole grains.

Some of the most important vitamins for pygmy goats include Vitamin A and Vitamin D. Vitamin A is essential for keeping the skin and organs healthy. This vitamin can be found in green, leafy hay as well as yellow corn. These materials all contain carotene, a substance which the goat's body will convert into Vitamin A. Vitamin D is essential for the proper use of phosphorus and calcium in building and repairing bones. During warm weather, pygmy goats receive their required Vitamin D through natural sunlight. In the winter, however, it may be necessary to provide sun-cured hay.

The minerals most required by pygmy goats include calcium, phosphorus, iodine and selenium. Feeding them alfalfa hay is a great way to provide goats with calcium while whole grains are a good source of phosphorus. Though calcium and phosphorus are the most important, it may be beneficial to supplement your goats' diets with iodine and selenium as well. Selenium is known to prevent white muscle disease in goats and iodine can be given by offering iodized salt.

a.) Vitamin and Mineral Supplements

In addition to providing your pygmy goats with a well-balanced diet of grass, hay and grain, you may also want to think about incorporating some vitamin and mineral supplements. The vitamins and minerals your pygmy goats need may vary according to the region in which you live, so be sure to contact a local veterinarian for recommendations. The reason for this is that the type of food available may be different in different regions, thus affecting the need for dietary supplementation.

Supplementation is most commonly used by pygmy goat breeders and those who produce pygmy goats for show. There is a great deal of pressure to raise well-developed kids that grow quickly. Many breeders use vitamin and mineral supplementation with varying degrees of success but it is not something the average goat owner should take lightly. Supplementation can be tricky for pygmy goats and, unless you understand how to do it properly, you could actually end up doing harm to your goats.

Pygmy goats that are offered a well-balanced diet generally do not require much supplementation. They may benefit from the addition of trace mineral salt to their diet as well as any nutrients that are lacking in your specific region. It is important to note that the supplements designed for other livestock including cattle and horses should not be used for pygmy goats. Pygmy goats have a very low tolerance for copper than other livestock and livestock supplements often contain more copper than a pygmy goat's body can handle.

It is also possible for your pygmy goat to overdose on certain fat-soluble vitamins including Vitamins A, D and E. These vitamins may be stored in the body if the body's requirements are exceeded, but prolonged storage of these vitamins in excess of the body's requirement can lead to toxicity in pygmy goats. If you do choose to supplement the diet of your pygmy goats, be sure to read the warning labels on the product so you only give your goats the amount they need – even the smallest bit extra has the potential to cause health problems down the line.

In most cases, supplementation is not necessary for pygmy goats. If you allow your kids to nurse for 10 to 12 weeks and offer them a healthy diet, they should not need any supplementation in order to achieve proper growth. If you over-supplement the diet of your kids, they may end up having health problems later on in their lives.

b.) Feeding Summary

Stomach Type: four compartments: rumen, reticulum, omasum, and abomasum
Begin Nursing: immediately after birth
Fully Weaned: 8 to 10 weeks
Feeding Method: grazing/foraging
Preferred Foods: broad-leafed plants, weeds, legume hay
Protein: alfalfa and clover hay, soybean meal
Carbohydrates: whole or rolled grains
Vitamins: Vitamin A from leafy hay and yellow corn, Vitamin D from sun exposure
Minerals: calcium (whole grains), phosphorus (alfalfa hay), iodine (iodized salt) and selenium

4.) Building a Playground for Pygmy Goats

Keeping pygmy goats as pets can be an enjoyable and rewarding experience. These animals can be very entertaining to watch, especially if you provide them with a playground. A playground for pygmy goats does not necessarily need to be complex or expensive – in fact, you can probably build one using recycled materials you have around the house. Providing your pygmy goats with a playground will help them to stay active and to form bonds with other goats in the herd.

In order to build a playground for pygmy goats, all you really have to do is assemble a collection of objects your goats can climb on, jump over or race through. You might, for example, create a maze out of upright tires or arrange tree stumps of varying size in a line so your pygmy goats can hop from one to another. Be creative when designing your pygmy goat playground and don't be afraid to rearrange it once your pygmy goats spend a little time in it and you are able to get a feel for what they like.

In addition to considering what type of objects to include in your pygmy goat playground, you should also think about where you want to place it. If you have a very large fenced area for your pygmy goats to roam in, dedicate a certain portion of the area to create your playground. You can also fence off a separate portion of your yard to use as a playground for pygmy goat kids. The ideal location for a pygmy goat playground will be such that it catches the sunlight in the morning but is shaded by trees or shelter in the afternoon.

<u>Objects to use in Playgrounds</u>:

- Old tires
- Tree stumps and logs
- Wooden ramps
- Old farm equipment
- Large rocks
- Mounds of dirt
- Wooden see-saw

The options are endless when it comes to building a playground for your pygmy goats so don't be afraid to get creative!

Chapter Six: Keeping Your Pygmy Goats Healthy

If you plan to keep pygmy goats as pets it is your responsibility to keep them as healthy as possible. Not only does this include providing adequate food and shelter, but you will also be required to provide vaccinations and access to veterinary care. No matter how careful you are, your goats are likely to be exposed to disease at some point during their lives. In these situations it is imperative that you take steps to identify the disease quickly and to seek treatment.

Before you bring your pygmy goats home, it would be wise to familiarize yourself with some of the most common health problems affecting pygmy goats so you know what to look for. If you are prepared with knowledge about common pygmy goat health problems, you will be better equipped to handle them as they arise.

1.) Normal Health Values for Pygmy Goats

The key to ensuring that your pygmy goats remain healthy is to know enough about them to realize when they are sick. Though learning the signs and symptoms of common pygmy goat diseases is certainly important, you cannot truly know when your pygmy goats are sick unless you know what they are like when they are healthy.

The following values are normal for healthy pygmy goats:

Rectal Temperature - 102.5 - 104°F
Pulse Rate - 60 to 80 beats per minute

Respiration Rate - 15 to 30 breaths per minute

It is important to note that the average values may vary from one goat to another but any value within the ranges listed above can be considered normal. If your pygmy goat is nervous, he may have a higher pulse or respiration rate. If you suspect this to be the case, make allowances for variables in your measurements.

To measure your goat's respiration rate, simply watch the movement of its ribcage, counting the number of breaths the goat takes in one minute. In order to measure a pygmy goat's pulse rate, you may need to use a stethoscope. This method may require you to enlist help in holding the goat still while you take the pulse rate. Another option is to feel for the goat's pulse with your fingers – position your fingers on the front of the goat's chest near the bottom, just over the heart.

In addition to these health values, it is also a good idea to familiarize yourself with normal values for various stages in life. Some of the key life stages to know for your pygmy goats include the duration of the estrous cycle and gestation. It is also helpful to know the age at which your goats are likely to hit puberty.

The following values are normal for healthy pygmy goats:

Estrous Cycle - 18 to 23 days
Estrum - 12 to 36 hours
Gestation - 145 to 153 days
Puberty - 4 to 12 months old

Puberty is simply the age at which pygmy goats become sexually mature – for most goats, this falls between the age of 4 and 12 months. Sexual maturity simply means that a doe is capable of being impregnated and a buck is capable of impregnating a doe. It is possible for a doe to become pregnant as early as 2 months of age, however, so it is best to separate the sexes before the kids reach 2 months old.

Female pygmy goats go into season (called estrum) every 18 to 23 days, on average. The interval of time in which the doe is in estrum (ready to receive a buck for breeding) generally lasts between 12 and 36 hours. After estrum has begun, the female goat typically ovulates within 24 to 36 hours.

If the doe is impregnated, gestation of the kid typically lasts between 145 and 153 days. If the kid is born prior to the 145-day mark, it is likely to be premature and has a lesser chance of survival. Kids born prior to the 139-day mark are unlikely to survive more than a few hours.

2.) Common Health Problems

Enterotoxemia

This disease is caused by bacteria called *Clostridium perfringins*. These bacteria are naturally found in the gut of pygmy goats but, under certain circumstances, they can reproduce to toxic levels. There are several different types of this disease and it can affect pygmy goats at various stages in life. Type C, referred to as bloody scours, typically affects kids during the first few weeks after birth. The onset of the disease is usually precipitated by an increase in feeding. Type D, called overeating disease, affects kids over one month of age and it often occurs following a sudden change in feed. Treatment for this disease is often ineffective but vaccination can help to prevent it.

Cause: *Clostridium perfringins* bacteria

Symptoms: loss of appetite, lethargy, diarrhea, stomach pain

Treatment: treatment is ineffective, vaccination recommended

Tetanus

This disease is caused by bacteria called *Clostridium tetani*, which may enter the body through open wounds or broken skin. Once it enters the body, the bacteria spread and reproduce to toxic levels. Also called lockjaw, tetanus typically causes muscle spasms and may cause the body of the affected goat to become stiff and rigid. Treatment is typically ineffective and the condition is often fatal – it can, however, be prevented with vaccination.

Cause: *Clostridium tetani* bacteria

Symptoms: muscle spasms, rigid gait, bloat, anxiety, inability to open mouth, excess salivation

Treatment: treatment is ineffective, vaccination recommended

Rabies

This disease is more common in wildlife than in livestock, but it can still be spread to pygmy goats through a bite from an infected animal. Though there isn't a rabies vaccine designed specifically for goats, sheep vaccines have proven effective in multiple cases. Unfortunately, vaccination after a goat has been bitten will have no effect on the disease. Because the cost of these vaccinations is high and the disease quite rare, it is up to the owner to decide whether it is warranted or not.

Cause: bite from an infected animal

Symptoms: aggression, apprehension, hyperactivity, loss of appetite, excess salivation, frothing at the mouth

41

Treatment: treatment is ineffective, vaccination recommended

Johne's Disease

This disease is a very serious condition that affects a variety of ruminants including goats, sheep and cattle. Johne's disease is caused by the bacteria *Mycobacterium avium* subsp. *pseudoparatuberculosis*, which can survive in soil for long periods of time. For this reason, there is often a long delay between the time of infection and the manifestation of symptoms. This disease tends to spread very quickly undetected which can result in the loss of the entire flock. Kids are most susceptible to this disease and they can catch it from an infected doe. The best way to prevent this disease is to engage in proper sanitation and regular testing of the herd.

Cause: *Mycobacterium avium* subsp. *pseudoparatuberculosis* bacteria

Symptoms: symptoms may not appear until the late stages of the disease; rapid weight loss, diarrhea, weakness

Treatment: proper cleaning and sanitation, regular testing

Sore Mouth

A type of viral skin disease, sore mouth is also called contagious ecthyma. There is a vaccine for this disease but administration can be dangerous. The vaccine for this disease is "live" which causes the vaccinated animal to develop lesions at the vaccinated location. These sores are highly contagious for humans so it is important to wear gloves during administration. Unless sore mouth is a problem among your herd, it is best not to vaccinate for it because the vaccination will introduce the virus to your herd. Once you begin vaccinating for sore mouth, you need to re-administer the vaccine annually.

Cause: skin virus, spread through contact

Symptoms: lesions, pustules, scabs

Treatment: treatment is often ineffective, vaccination recommended

Dermatitis

In simple terms, dermatitis is inflammation or irritation of the skin. Dermatitis can take many forms in pygmy goats and it may have several different causes. In many cases, dermatitis presents itself as a rash but it may also produce blisters, scaling or crusting. There are seven different forms of dermatitis affecting goats: staphylococcal, labial, interdigital, pustular, malasezzia, alopecic exfoliative and herpetiformis.

Dermatitis can be caused by a number of things including exposure to toxic substances, inhaling chemicals or detergents and dryness of the skin. Fungal infections can also result in dermatitis, as can exposure to allergens in feed or hay. Eating certain plains such as nettles or azaleas may expose the goat to mites that can cause dermatitis – bites from mosquitoes or wasps may also cause skin problems. Treatment for dermatitis varies depending on the cause of the condition but a number of creams and ointments including lanoline, petroleum jelly, betamethasone cream and diazinon.

Cause: exposure to toxic substances, inhaling chemicals or detergents, dryness of the skin, fungal infections, exposure to allergens in feed or hay, insect bites

Symptoms: itchy red rash, swollen bumps, boils or pustules, scaling of the skin, blisters

Treatment: application of creams and ointments including lanoline, petroleum jelly, betamethasone cream and diazinon.

Parasites

Due to their grazing behavior, pygmy goats are highly susceptible to developing worms. Though all goats have worms to some degree, those that cause the most damage are coccidian and stomach worms. The most dangerous stomach worm is the barber pole worm, *Haemonchus contortis*, which pierces the stomach lining, causing blood loss and anemia. These worms have a very short life cycle, which makes them very difficult to control – they also reproduce very quickly and can go into hibernation if necessary for survival. Another type of stomach worm is the brown stomach worm, *Ostertagia cicumctinca*, which also burrows into the stomach lining. This worm typically causes stomach upset and diarrhea.

Coccidia are also commonly found in goats. A type of single-cell protozoa, coccidia are known for damaging the lining of the small intestine where nutrient absorption takes place. This often causes stunted growth in kids as well as weight loss and chronic diarrhea. Whereas stomach worms are often transmitted through grazing, coccidia tends to affect herds that are in confinement – particularly when sanitation is poor.

Urinary Calculi

Also known as "water belly," urinary calculi is a form of urinary tract disease that is fairly common in goats. This disease is most common in males and it can prevent both urination and breeding. Female goats are less likely to contract this disease because their urethras are shorter and straighter. The longer, less straight urethra of male goats makes it more difficult to pass solid particles. When solid particles collect in the urethra, it results in urinary calculi.

This condition is often the result of improper feeding. If you feed your goats an improper ratio of calcium to phosphorus this can result in the build-up of solid particles in the urine. The ideal ratio of calcium to phosphorus in a pygmy goat's diet is 2 ½ to 1. Overfeeding of grain concentrates is also a common cause of this condition as are certain types of hay that are high in phosphorus and hay fertilized with chicken litter.

The accumulation of solid particles in the urine may block the flow of urine out of the goat's body. As a result, the goat may experience pain or discomfort in urination – in extreme cases, this condition can also lead to death. This condition will not go away on its own and it can quickly become fatal if not properly treated. It may be necessary to correct the problem surgically, though treatment with ammonium chloride may also be effective. It is important that you do not give goats suffering from this condition too much water because they will be unable to expel it.

Cause: collection of solid particles in the urethra; often the result of improper feeding

Symptoms: restlessness, anxiety, straining to urinate, bloated body, pain or discomfort in urination

Treatment: surgery may be necessary in extreme cases; ammonium chloride treatment may also be effective.

Pneumonia

The term pneumonia is used to describe inflammation or infection of the lungs in goats. Other infections like bronchitis, tracheitis and laryngitis are generally limited to the upper portion of the respiratory tract. Pneumonia is a very serious disease and it is often due to poor management. Environmental factors such as poor ventilation, overcrowding and dirty conditions greatly contribute to the risk of pneumonia.

Other contributing factors for pneumonia in goats include lungworms, stress, aspiration of toxic materials and exposure to virus. Lungworms may not cause pneumonia directly, but they can damage lung tissue, which leaves the goat more susceptible to secondary infection. Stress can also lower a goat's resistance to disease – this is particularly common in show goats that are exposed to the stress of shipping and showing. Aspirating toxic materials such as drenching substances can also lead to respiratory infection.

Aside from these factors, some kind of virus, bacteria or fungus may also be an infection agent for pneumonia. Some of the bacteria most commonly responsible for pneumonia include Pasteurella and Corynebacterium. Infection is often spread through the air and it tends to spread more quickly in overcrowded conditions and when new animals are introduced into an existing herd.

Some of the most common symptoms of pneumonia include labored breathing, rapid breathing, a rattle in the chest, intolerance to exercise and nasal discharge. Goats suffering from pneumonia may eventually lose weight and, in extreme cases, may also die. In the initial stages of the disease, a fever may be present but the animal's temperature is generally normal throughout the course of the disease. Some goats develop a chronic cough, but this is typically a sign of bronchial or laryngeal infections.

Treatment for pneumonia involves identifying the underlying cause of the infection and providing a remedy. In the case of lungworms, your veterinarian may take a fecal exam or he may take a culture of nasal discharge. It is important to isolate goats suffering from pneumonia to prevent the spread of the disease. It is also important to engage in proper vaccination and sanitation procedures to prevent disease in the first place.

Cause: may be caused by virus, bacteria or fungus; can also be the result of weakened immunity due to overcrowding, poor ventilation or stress

Symptoms: labored breathing, rapid breathing, chronic cough, fever, nasal discharge, rattle in the chest, intolerance to exercise

Treatment: identification and remedy of the underlying cause for the infection

3.) Treating Pygmy Goat Illness

When it comes to treating illness in your pygmy goats it is always wise to seek the guidance of a veterinarian. Your veterinarian will be able to help you correctly identify the disease and will also recommend the proper treatment. If medication is necessary, one of three types of injection may be used:

Intramuscular (IM): This type of injection is typically given in the neck muscle behind the head and it should be injected slowly. Because these injections can cause muscle damage, they should never be given in the loin or leg regions.

Intravenous (IV): These injections are given directly into the vein and they are typically used when it is necessary to get the medication directly into the blood stream to facilitate a quick response.

Subcutaneous (SQ): This type of injection is given under the skin, typically behind the point of the shoulder. To give an SQ injection you must make a "tent" with the skin by pinching it and pulling it gently away from the muscle. The needle can then be inserted and the medication injected into the pocket beneath the skin. These injections are most effective when administered slowly.

Oral Medications: The process of administering oral medications to goats is also referred to as drenching. Anti-parasitic drugs should always be given orally, even when an injectable product is available. To administer oral medications, you will need a drenching gun – these are available in single- and multi-dose form. To use a drenching gun, insert the tip into the corner of the goat's mouth while gently restraining it. Push the plunger on the gun slowly so the medication passes over the goat's tongue. Once the goat has swallowed the medication, the gun can be removed and the goat released.

4.) Vaccinations

In addition to familiarizing yourself with the common health problems affecting pygmy goats, you should also be aware that pygmy goats require certain vaccinations. All of your goats should be vaccinated for clostridial diseases (particularly enterotoxemia) and tetanus. You may also want to consider vaccinating for sore mouth, caseous lymphadentitis and rabies – these vaccinations are only necessary in cases where the condition has been diagnosed or the risk is considered to be high. Make sure your vet will administer the injections, don't attempt to do it yourself, unless you are properly educated to do so.

The purpose of administering vaccines is to stimulate the immune systems of your goats to produce antibodies that will protect against the disease being vaccinated for. The vaccine itself will not protect your goats against disease and no vaccine is guaranteed to be 100% successful. After exposure to the vaccine, a healthy goat's immune system will produce antibodies against the disease over a period of about 4 weeks. After that time, the antibodies will generally provide long-term protection but some vaccines require occasional boostering.

When kids are born they have virtually no immune system – they receive their first antibodies from their mother's milk. By 10 weeks of age, the kid's immune system will begin to develop and that is the best time to start vaccinations. Vaccinations administered before this time have been shown to produce little measurable result.

Summary of Recommended Vaccinations:

Minimum Age - 10 to 12 weeks old
Recommended Vaccines - clostridial diseases (particularly enterotoxemia) and tetanus
Optional Vaccines - sore mouth, caseous lymphadentitis and rabies
Protection Period - protection is at its highest level about 4 weeks after exposure to the vaccine

5.) First Aid Kits for Pygmy Goats

If you plan to keep pygmy goats as pets, you should take the time to stock a first aid kit. Throughout the course of their lives, your pygmy goats will likely be exposed to minor injuries. These injuries may be the result of rough play or they could simply be accidents that happen in day-to-day life. Because veterinary care can be expensive, it would be wise to familiarize yourself with the basics of pygmy goat first aid and to keep a stocked first aid kit on hand in case you need it.

Recommended Supplies:

- Antibacterial ointment
- Antibiotic powder
- Aspirin
- Gauze wrap and pads
- Epinephrine
- Milk of Magnesia
- Penicillin injections
- Tetanus antitoxin
- Sterile eye wash

a.) Dealing with Common Injuries

Pygmy goats have a tendency to get a little rambunctious and it is not uncommon for them to experience minor injuries through the course of play. If you have a first aid kid on hand and are familiar with the proper method of dealing with these injuries, you can save yourself a lot of money on the cost of veterinary care – especially if you have multiple pygmy goats.

Cuts or Abrasions

Cuts and abrasions are fairly common in pygmy goats and, in many cases, they are very minor. If the cut is very shallow and there is little bleeding, all you need to do is clean the cut, apply an antibiotic ointment and cover the wound. For deep cuts, those that go through the skin with separated edges, you may need to suture the wound. If you are not qualified to provide this care yourself, you can clean the wound, apply antibiotic ointment and bandage the wound until you can get the goat to a veterinarian. Wounds that are very severe and/or spurting blood should be treated by a veterinarian immediately.

Another common injury involving cuts or abrasions may occur in trimming your goat's hooves. If you trim the hooves too far, you could cut the quick, which will result in profuse bleeding. In many cases, the bleeding stops on its own but if the cut is deep, you may need to wrap it and apply pressure to stop the bleeding. These wounds generally do not get infected but, if the cut is very deep, your vet may want to administer a penicillin injection just as a precaution.

Eye Injuries

It is important to treat eye injuries and abnormalities quickly
because even a small problem can progress into a serious one.
Pygmy goats may experience eye injuries or irritation from weeds
or straw becoming lodged under the eyelid. If not treated
promptly, this can lead to corneal lacerations and, potentially, loss
of vision. Common symptoms of this problem include squinting
and excessive tearing as well as eye closure and accumulation of
pus. To treat this type of injury, flush the goat's eye with sterile
eyewash and carefully remove the foreign body. You should also
treat the eye with antibiotic ointment for several days.

Leg Injuries

If you notice one of your goats limping, it is most likely due to
some kind of injury. Limping is a sign that the limb is causing the
goat pain so you should examine it for swelling and wounds. If
you are able to locate a physical wound, treat the goat with a dose
of penicillin and re-check the injury after 12 hours. In the case of
a strained muscle, treat the goat with aspirin at a rate of 5 grams
per 60 lbs. body weight and check the goat after 24 hours. In the
event that your goat is unable to move the leg or put any weight
on it, it is possible that the leg is fractured. This type of injury
needs to be treated by a veterinarian and should be supported until
the goat can be seen by a vet.

Digestive Problems

The most telling sign of digestive problems in pygmy goats is a
change in the stool. Pygmy goat stool is typically expelled in the
form of pellets that range from 0.5 to 1.5 cm in diameter. Dog-like
or watery stool may be an indication of diarrhea or a serious

digestive complication. In adult pygmy goats, watery stool may be an indication of parasitism, enterotoxemia or Johne's disease. A good product to treat diarrhea in pygmy goats is Oral sulfa – this is a common treatment for coccidiosis and bacterial bowel infections. If, after treating the condition, your pygmy goat shows no signs of improvement, it is best to seek veterinary attention to determine the cause of the digestive distress.

b.) Summary of Pygmy Goat First Aid

Recommended Supplies - Antibacterial ointment, aspirin, gauze, epinephrine, penicillin, sterile eye wash

Cuts and Abrasions - clean the cut and apply antibiotic ointment and a gauze bandage; deeper cuts may require sutures and penicillin injection to prevent infection. Always ask your vet to give injections, don't attempt to do it yourself, unless your vet told you to do so and has educated you how to do it.

Eye Injuries - symptoms include watering, eye closure, squinting, pus accumulation; most often caused by foreign bodies under the lid; flush with sterile eye wash and treat with antibiotic ointment after removing foreign body

Leg Injuries - examine leg for wounds or swelling; treat wounds with penicillin and swelling with aspirin; re-check the leg after 12 to 24 hours

Digestive Problems - may be caused by parasitism, enterotoxemia or Johne's disease; identified by change in stool and treated with oral sulfa

6.) Cleaning Up After Your Pygmy Goats

Pygmy goats are fairly clean animals by nature so cleaning up after them is not a significant concern. If you are able to keep your goats in a pasture, you will only have to worry about cleaning up after them inside the barn. Because pygmy goats are clean animals, they do not tend to roll in mud or sleep on dirty bedding. For this reason, you will need to keep a large supply of fresh hay on hand to refresh the bedding in your goat stalls.

If you do not have a pasture for your goats, you can use fencing to prevent them from wandering into areas where you do not want them to leave droppings. To preserve the cleanliness of the exercise pen, you might also consider keeping your goats inside the barn at night. In areas where the climate is very hot, make sure your barn has excellent ventilation so your goats do not suffer from heat stroke after being kept indoors for extended periods of time.

7.) Poisonous Plants/Substances

Whether you keep your pygmy goats in an enclosed area or in a pasture, it is important to make sure they do not come into contact with poisonous plants or substances. Pygmy goats do not know the difference between the plants that are safe for them to eat and those that aren't, so it is your responsibility to protect them. In the event that your goat does become poisoned, immediate veterinary care is essential. In extreme cases, fatalities occur despite treatment but providing your goats with immediate veterinary care will increase their likelihood of recovery.

It is important to note that well fed goats are less susceptible to poison. If you keep your goats well fed on healthy foods, they are less likely to forage and feed on potentially poisonous plants. When your goats are hungry, they may be less scrupulous about the things they eat and could also have a lower resistance to poison. The best preventive measure you can take against poison is to keep your pygmy goats well fed.

The following plants are considered poisonous to pygmy goats:

Avocado	Lilacs
Azalea	Lily of the Valley
Boxwood	Lupine
Cassava	Milkweed
Choke Cherries	Monkhood
Datura	Mountain Laurel
False Tansy	Nightshade
Fusha	Oleander
Holly	Red Maples
Japanese Pieris	Rhododendron
Japanese Yew	Rhubarb
Larkspur	Wild Cherry

In addition to plants, you also should be aware that certain substances can be poisonous to goats. Fertilizers, herbicides and other chemicals are extremely toxic to goats and should be kept out of reach. The following items should be considered toxic to pygmy goats:

- Fertilizers
- Gasoline
- Herbicides

- Insecticides
- Lead Paint
- Rodent Poison
- Cleaning Products

To keep your pygmy goats safe, it is wise to familiarize yourself with the symptoms of poisoning. These symptoms may vary in severity depending on the amount of the substance ingested. Remember, the sooner you are able to diagnose and treat the poisoning, the more likely your goats are to recover.

Common symptoms of poisoning include:

Bloating	Muscle Weakness
Colic	Excess Salivation
Coma	Staggering
Constipation	Vomiting
Convulsions	Weak Pulse
Death	
Dermatitis	
Diarrhea	
Dilated Pupils	
Fever	
Sensitivity to Light	
Frothing Mouth	
Hyperactivity	
Lameness	
Labored Breathing	
Muscle Spasms	
Rapid Pulse	

If you suspect that your pygmy goats have been poisoned, you should take the following steps immediately:

1. Take steps to prevent further exposure to the toxin.
2. Isolate the affected goat.
3. Provide plenty of fresh water and avoid stressing the goat.
4. Consult the packaging for suspected toxic material to confirm diagnosis of poisoning.
5. Contact your veterinarian immediately for treatment.

The following plants are safe for pygmy goats to consume:

Acorns	Ginger Root
Althea	Ficus
Apples	Fava Beans
Bamboo	Elm
Banana	Dandelion
Beans	Ferns
Beets	Hay
Blackberry Bushes	Hibiscus
Bramble	Honeysuckle
Broccoli	Ivy
Cantaloupe	Japanese Elm
Collard Greens	Johoba
Catnip	Mango Leaves
Carrots	Kudzu
Celery	Mesquite
Clover	Mint
Corn Husks	Mountain Ash
Cottonwood	Morning Glory
Citrus	Moss
Grapefruit	Mustard
Garlic	Nettles

Pea Pods

Onion

Pepper Plants

Pomegranates

Pumpkin

Potatoes

Raspberry Plants

Sunflowers

Turnips

Yarrow

Virginia Creeper

Watermelon

Wandering Jew

Weeping Willow

Wild Rose

Wild To

Chapter Seven: Showing Pygmy Goats

Showing pygmy goats can be an incredibly rewarding experience but it can also be a challenge. As is true of dog shows, pygmy goat shows require the goats to meet certain standards for conformation. Many pygmy goat breeders breed their goats in order to achieve these standards, thus improving their likelihood of winning a show.

If you are interested in showing your pygmy goats but have no experience in the area, start off by attending a few Agricultural County Shows where you are likely to find a variety of livestock shows including those for pygmy goats. There are a few general things you should do or look for in preparing your pygmy goat for show.

Included in this list are:

1. Checking to be sure your goat does not display any disqualifying faults.
2. Teaching your goat to walk on a lead and to stand quietly.
3. Accustom your goat to being touched and to having its mouth opened for inspection.
4. Brush your goat on a daily basis for several weeks before the show.
5. Trim your goat's hooves and bathe your goat two days prior to the show.
6. Rub your goat's horns and hooves with a little bit of hoof oil to give them shine.

7. Assemble the supplies you need to take with you to the show: water bucket, hay, brush, collar and lead.
8. Review the rules and directions for the show and follow the instructions of the steward or judge.

1.) Pygmy Goat Breed Standard

If you seriously intend to show your pygmy goats, you will need to cultivate a firm understanding of the pygmy goat breed standard. The breed standard provides guidelines for size, color, coat and composition of male and female pygmy goats. There are separate categories for males, females and wethers and each category has a number of identified disqualifying faults.

Breed Standard Basics

<u>**Size**</u>

- Females should exhibit a minimum size of 17" at the withers and a maximum size of 21"
- Males should exhibit a minimum size of 17" at the withers and a maximum size of 22"
- Cannon length (measured from the knee and pastern joint to the cannon bone) should be a minimum of 3 ½" and a maximum of 4 ½"

<u>**Composition**</u>

- The head should be medium long with a straight or dished profile
- The muzzle should be rounded with a full chin
- The forehead should be broad and flat to slightly concave

- The eyes should be wide-set, bright and darkly colored
- The ears should be medium-sized and firm but mobile
- The body should be large in proportion to the size of the animal
- The chest should be broad and deep, increasing in width towards the flank but maintaining symmetry
- The back should be strong and laterally straight, sloping down from the withers into the loin
- The rump should have a gradual slope with a pronounced, high-set tail
- The hips should be wide and level with the back
- The legs should be strong and well-muscled
- The forelegs should be short and straight, squarely set with the elbows close to the ribs
- The hind legs should be straight and widely set

Coat

- Pygmy goats should exhibit a full coat of straight hairs
- Coat may vary in length and density
- Females may exhibit sparse beards or no beard at all – they should not exhibit trimmed beards
- Males should exhibit a full, long beard with a mane over the shoulders

Color

- All colors and markings are acceptable with the exception of "Swiss stripes" on the face

Sex-Specific Standards

- Females should exhibit a firm, rounded udder with symmetrically placed teats
- Males should exhibit longer, more substantial horns than females, though disbudding is permissible
- Males should exhibit two testicles of appropriate size carried in a healthy scrotum
- Wethers should exhibit slightly less horn growth than intact males and do not develop the same cape-like mane

Disqualifying Faults

- Mouth and jaw defects (over-shot, under-shot or twisted)
- Teat defects in males or females
- Genetically hornless (polled)
- Roman nose or pendulous ears
- Not meeting size standards
- In males, testicles not properly sized or descended
- Supernumerary teats removed

Pygmy goats that exhibit certain disqualifying faults can still be registered in the Pet Record if not the Herd Book. This record is open to females and wethers born to unregistered parents as well as those exhibiting mouth, teat or size faults.

2.) Scoring for Pygmy Goat Shows

Upon entering your pygmy goat in a show, it will be judged based on its conformance to a number of categories. Different scorecards are used for does, bucks and wethers, each totaling 100 points. There is also a 100-point scorecard used for showmanship – this is used to evaluate the prowess of the showman and it is scored separately from the goat itself.

a.) Scoring for Pygmy Goat Shows

General Appearance (Doe 14 pts, Buck 14 pts)
- Body measurements meet the age group standard
- Genetically horned
- Appearance is balanced and styled
- Proportions are wide in relation to height and length
- Health is perfect and condition optimal

Head and Expression (Doe 10 pts, Buck 12 pts)
- Head is medium-short, dished profile
- Jaws are broad and strong, symmetrical
- Bite is even
- Eyes are bright and widely set
- Ears are medium-sized, firm and erect
- Muzzle is broad and full
- Nose is short and flat
- Expression is animated and alert

Coat (Doe 4 pts, Buck 6 pts)
- Coat is dense; abundant in bucks
- Hair is straight and medium long

Breed Markings (Doe 8 pts, Buck 12 pts)
- Distinct breed-specific markings
- Light color accents on muzzle, forehead and eyes

- Contrasting darker color on crown, cannons and legs
- Light girth areas are acceptable

Neck (Doe 3 pts, Buck 5pts)
- Neck is strong and muscular
- Full-throated, blends into the withers

Shoulders (Doe 5 pts, Buck 5 pts)
- Shoulders are angulated and laid back
- Shoulder blades are firmly attached
- Withers are nearly level with the spine

Chest (Doe 10 pts, Buck 10 pts)
- Chest is floor wide, prominent fore chest
- Heart girth is large and full at the elbows
- Ribs are well sprung, long and set wide apart

Barrel (Doe 8pts, Buck 8pts)
- Barrel is symmetrical, broad and deep
- Barrel widens toward the flanks

Back (Doe 8 pts, Buck 8 pts)
- Back is strong and broad
- Back is straight and level long the chin and loins

Rump (Doe 8pts, Buck 8pts)
- Rump is medium-long and medium-wide
- Hips are wide and nearly level with the spine
- Thurls are set high and wide apart
- Pinbones are prominent and set well apart
- Tail is symmetrical and carried high

Legs and Feet (Doe 10 pts, Buck 12 pts)
- Legs are strong and muscular, squarely set
- Forelegs are straight and wide apart
- Cannon bone is short, elbows set close to ribs
- Hind legs are short-hocked and angulated; aligned with hips
- Pasterns are short and strong
- Feet are well-shaped and symmetrical
- Heels are deep with level soles
- Gait is smooth and balanced, effortless

Mammary System (Does 12 pts)
- Teats are cylindrical and symmetrical in shape and placement
- Teats are functional
- Teats are free from deformity and obstruction
- Udder is function, firm and balanced
- Udder is small to medium sized, well-attached

Reproductive System (Bucks, no pts)
- Testicles are of normal size
- Testicles are equal in size and fully descended

Mammary System (Bucks, no pts)
- Teats are normal and non-functional
- Two single teats, free from deformity

Points are awarded in the categories above based on the goat's conformity to the breed standard. Faults are evaluated and rated from moderate to very serious. Judges use a detailed scoring sheet to evaluate the presence of disqualifying faults or minor faults that result only in a reduction of points.

b.) Summary of Show Scoring

General Appearance - Doe 14 pts, Buck 14 pts
Head and Expression - Doe 10 pts, Buck 12 pts
Coat - Doe 4 pts, Buck 6 pts
Breed Markings - Doe 8 pts, Buck 12 pts
Neck - Doe 3 pts, Buck 5pts
Shoulders - Doe 5 pts, Buck 5 pts
Chest - Doe 10 pts, Buck 10 pts
Barrel - Doe 8pts, Buck 8pts
Back - Doe 8 pts, Buck 8 pts
Rump - Doe 8pts, Buck 8pts
Legs and Feet - Doe 10 pts, Buck 12 pts
Mammary System - Does 12 pts, Bucks no pts
Reproductive System - Bucks no pts

c.) Scoring for Showmanship

The showmanship of pygmy goat showmen is also evaluated on a scale of 1 to 100. Showmen are evaluated in three different categories for a total of 100 points.

Appearance of the Animal (20 possible points)
- Condition and general appearance (goat is not too fat or thin, well groomed and ready for show) – **5 points**
- Hoofs are trimmed and shaped – **5 points**
- Cleanliness (animal is clean and free from stains, groomed for show) – **10 points**

Appearance of the Exhibitor (10 possible points)
- Clothes and person are neat and presentable
- Proper attire for show ring (sandals, hats and shorts not allowed)

Showing the Animal in the Ring (70 possible points)
- Leading the animal – **10 points**
  - ○ Entering at a normal pace and walking on the left side around the ring in a clockwise direction
  - ○ Goat responds quickly and leads readily
  - ○ Proper lead equipment
  - ○ Goat's head is held high, walk is graceful
  - ○ Proper maneuvers are exhibited upon request
- Setting up the animal – **15 points**
  - ○ Setting up the animal to its best advantage
  - ○ Standing where both the judge and goat can be observed
  - ○ Setting up the goat so the front feet are square and the hind feet slightly spread
- Recognizing conformation faults and striving to overcome them appropriately – **5 points**
- Maneuvers – **20 points**
  - ○ Leading the goat properly according to the rules
  - ○ Responding correctly to the judge's requests
  - ○ Poised and alert attitude in the ring
  - ○ Courteous and sportsmanlike behavior at all times
- Knowledge of the breed – **20 points**
  - ○ Properly answers questions about pygmy goats
  - ○ Answers correctly questions about goat terminology

Other Considerations for Showmanship

- Being polite to judges and other exhibitors
- Leaving an appropriate amount of space when walking or setting up goats in line
- Carefully listening to the directions of the judge
- Having poise and confidence during the show
- Not allowing the goat's knees to touch the ground
- Exhibiting a goat that has been properly groomed and cleaned
- Maintaining eye contact with the judge
- Standing on the correct side of the goat during show
- Keeping the goat between the exhibitor and the judge
- Having excellent knowledge about the pygmy breed
- Setting up the goat to its best advantage
- Consistently setting up the correct leg first
- Exhibiting maneuvers correctly and with fluidity

d.) Summary of Showmanship Scoring

Condition and General Appearance – 5 points
Hooves Trimmed and Shaped – 5 points
Cleanliness– 10 points
Appearance of the Exhibitor - 10 points
Leading the Animal – 10 points
Setting up the Animal – 15 points
Recognizing/Overcoming Conformation Faults – 5 points
Maneuvers – 20 points
Knowledge of the Breed – 20 points

Chapter Eight: Common Mistakes Owners Make

In this chapter you will find explanations of some of the most common mistakes pygmy goat owners make. If you want to avoid having problems with your goats, take the time to familiarize yourself with these mistakes so you can avoid making them with your goats.

Inadequate Fencing or Space

Pygmy goats may be small but they do need plenty of room to run and jump around. These small animals are also surprisingly good at jumping and climbing so if you choose a fencing material that is too short or not strong enough, your goats may be able to escape the enclosure. Make sure you provide a minimum of 15 square feet per pygmy goat and maintain a fence height around 4 feet off the ground.

Not Getting a Permit

If you have never kept livestock before, you may not realize that a permit is necessary in order to keep pygmy goats on your property. If you live in a residential area, you may be subject to certain zoning restrictions, which limit the number of pygmy goats you can keep or prohibit you from keeping them entirely. Check with your local council to find out what kind of restrictions are in place and to see what you need to do in order to obtain the proper license. It is important that you go through the licensing process before you actually purchase a goat to avoid the possibility of hefty fines later.

Being Unprepared for the Responsibility

Buying and keeping a pygmy goat is just as much work and responsibility as it is to keep a dog. Just because you keep your pygmy goat outside in an enclosure rather than in the house doesn't mean you don't still have to care for it. If you plan to buy a pygmy goat you should be prepared to provide it with adequate food and shelter – you will also need to provide routine veterinary care and all of the necessary vaccinations to keep your goat healthy. If you are not willing to commit to performing these responsibilities for the duration of your goat's life, do not purchase a pygmy goat.

Keeping Pygmy Goats Alone

Pygmy goats are very social creatures and they prefer to be kept in herds. If you don't intend to breed your goats, consider keeping two wethers (castrated males). It is also possible to keep a pygmy goat as a companion animal for horses or other livestock – they also get along surprisingly well with dogs, particularly Anatolian shepherds.

Keeping too Many Pygmy Goats

Though they are much smaller than most other livestock, you still need to be careful about how many pygmy goats you keep together. Refer to zoning requirements in your region regarding the maximum number of goats you are allowed to keep on your property and do not go over that number. Do not feel like you have to keep the maximum number, either – keeping fewer goats in a larger space is a much better option for the health of your goats than keeping too many goats in a small space.

Not Separating the Sexes

Pygmy goat nannies are capable of conceiving at a very early age so, if you do not take the precaution of separating males from females early on, you may end up with an unwanted pregnancy in your pygmy goat. Even if you do plan to breed your goats, it is best to keep the sexes separate until they are fully mature and then you should supervise breeding activities. It is not a good idea to keep intact males together with female goats because now only could this result in unwanted breeding, but it could also result in aggression between the males.

Overfeeding or Changing Diet

These creatures are notorious for being picky eaters – they also do not tolerate changes in their diet very well. Once you get into a routine for feeding your goats, don't make any sudden changes. Sudden dietary changes can cause extreme stomach upset in pygmy goats, and even dangerous conditions such as enterotoxemia. If you need to make changes to your goats' diet, do so gradually over a period of 7 to 10 days.

Improper Breeding Techniques

Though pygmy goats are prolific breeders by nature, that doesn't mean you don't need to do any research before you breed them.

71

Always take the time to make suitable matches for your does to ensure healthy, well-bred kids. Never mate a large-framed male to a moderate-framed female because this could result in birthing difficulty.

Weaning Kids Too Early

When pygmy goat kids are born, they immediately begin to nurse. Over the next few days, the kids may also begin to graze on roughage or even eat a little grain. Though they may eat these foods, the stomachs of pygmy goat kids are not yet developed enough to handle a diet composed solely of grains and vegetable matter. Adult pygmy goats have stomachs with four compartments but, when first born, the ruminant compartment of the stomach isn't fully developed. During the first few weeks of life, a pygmy goat kid's body acts like it has only one stomach – milk goes straight through to the appropriate compartment rather than passing first through the rumen.

After 6 to 8 weeks, the rumen should be sufficiently developed for your pygmy goat kids to properly digest plant matter. After about 10 weeks of age, it is appropriate to begin weaning them off their mother's milk. Though the kids may already be eating some solid food, they will likely continue to nurse if they are left with their mother. For this reason, it is best to separate the kids from the mother after 10 to 12 weeks of age.

Chapter Nine: Frequently Asked Questions

In this section you will find a number of frequently asked questions regarding pygmy goats. If you have questions about buying, feeding, breeding or caring for pygmy goats, look here to find the answer to your question.

The topics covered in this section include:

General Questions
Buying and Keeping Pygmy Goats
Breeding Pygmy Goats
Feeding Pygmy Goats
Pygmy Goat Health

General Questions

Q: Are pygmy goats different from domestic goats?

A: Pygmy goats are actually a breed of miniature domestic goat. Goats belong to the *Capra* genus, which includes up to 9 different species including the wild goat. Domestic goats are simply a subspecies of the wild goat and pygmy goats are a breed within that subspecies.

Q: How big do pygmy goats get?

A: The size of pygmy goats varies depending on sex and breeding, but the average size ranges between 50 and 90 lbs. with a height between 16 and 24 inches. Female goats (does) tend to be smaller, weighing from 50 to 75 lbs. while male goats (bucks) weigh between 60 and 90 lbs.

Q: Are there different species of pygmy goat?

A: Pygmy goats are not a separate species of goat but a breed within the domestic goat subspecies. There are not separate species or breeds of pygmy goats, but they do come in a wide variety of colors. Pygmy goats can be found in solid or multi-colored patterns including colors such as white, black, caramel, grey, brown and grizzled (agouti).

Q: Do pygmy goats make good pets?

A: Goats are kept for a number of reasons – they can be raised as meat animals, kept as companion animals for other livestock or used in research. Pygmy goats, however, are most popular as pets.

These animals are very gentle and affectionate by nature so they make excellent pets – particularly for children.

Q: What kind of climate is best for pygmy goats?

A: Pygmy goats are very hardy creatures so they adapt well to a variety of different climates - just be sure to provide your pygmy goat with the shelter it needs. If you live in a very hot climate, provide your goats with an open shelter to protect them from the sun. In cold climates, the shelter should be completely enclosed to keep your goats warm.

Q: How long do pygmy goats live?

A: The lifespan of pygmy goats varies depending on its diet and how well it is cared for. In general, however, pygmy goats live between 10 and 15 years.

Q: When did pygmy goats first become popular as pets?

A: Domestic goats originally come from West Africa and they were domesticated thousands of years ago – as early as 7,000 B.C. It wasn't until the 1950s, however, that domesticated pygmy goats were imported from Africa to Europe. Throughout the 1950s, pygmy goats were kept in zoos as exotic animals. At the end of the decade, however, shipments of these animals were sent to the United States where they became used as pets by private breeders.

Buying and Keeping Pygmy Goats

Q: Do I need a permit to keep pygmy goats?

A: Permits or licenses are generally required for keeping livestock and pygmy goats are no exception. If you plan to keep pygmy goats it would be wise to check with your local council regarding livestock zoning restrictions. The restrictions may vary depending on where you live, but some areas may limit the number of goats you can keep or set a minimum for the amount of space you need to provide. Licensing requirements may vary between the U.S. and the UK so it is best to research the requirements for your specific area before purchasing pygmy goats.

Q: How much space do I need for my pygmy goats?

A: If you perform some basic research regarding this question you will find a number of different answers. The minimum amount of space you should provide for pygmy goats is 15 to 20 square feet per goat. A pair of goats can safely be kept in a 30-by-30-foot pen, as long as they are given ample exercise space as well. Keep in mind that these recommendations are minimums – the more space you can provide for your pygmy goats, the better.

Q: Can I keep 1 pygmy goat by itself?

A: You can keep a single pygmy goat as a pet, but these animals tend to do better when they are kept in groups. Pygmy goats are very social animals and they prefer to be kept in herds. They can also be used as companion animals for horses and other livestock so if you only keep one pygmy goat, it would be wise to keep it with some other kind of animal for companionship.

Q: Is it difficult to keep pygmy goats?

A: Pygmy goats are actually fairly easy to keep and relatively inexpensive compared to other livestock. Because they are small, pygmy goats require less space and food than traditional livestock and they are generally easy to care for. In fact, pygmy goats that are properly raised around humans make great pets for children.

Q: What kind of restrictions are there for keeping pygmy goats?

A: Zoning restrictions vary from one region to another, so be sure to check with your local council for specific information regarding keeping pygmy goats on your property. You may be limited to a certain number of goats and you may also be required to provide certain enclosures or fencing to contain the goats. All of the restrictions should be outlined in the permit.

Q: Do I still need a permit in the UK?

A: UK permit requirements for keeping pygmy goats are different than in the US. You will need to apply for a holding register with the Department for Environmental Food and Rural Affairs and tag all of your goats electronically.

Q: Can I keep male and female goats together?

A: It is possible to keep males and females together as long as the males have been neutered. If the males are intact, it is best to house them separately from females to avoid unwanted breeding and aggression between males.

Q: What other animals can I keep with pygmy goats?

A: Pygmy goats make great companion animals for horses and other livestock – they are often used in horse breeding facilities and at racetracks. These animals also get along well with dogs and other household pets.

Q: How much does it cost to keep pygmy goats?

A: When it comes to figuring out costs you must consider both the initial and the recurring costs. To buy a pygmy goat you will need to first invest in the enclosure and then purchase the animal itself – these costs can range from $300 to over $1000 (from £185 to £620). You may also need to dehorn the goat and take it to the veterinarian in order to bring it up to date on vaccinations. Recurring costs for pygmy goats include food and veterinary care, which will vary depending on the number of goats you keep, and the type of food you choose to feed your goats.

Q: What are the pros of keeping pygmy goats?

A: Pygmy goats are much smaller than other livestock so they take up less space. These animals also make great pets and can even be used as therapy animals. Pygmy goats are easy to train and they can be bred all year-round.

Q: What are the cons of keeping pygmy goats?

A: In addition to needing a license to keep pygmy goats, you will also need to build a suitable enclosure – this can be very expensive. Pet goats may also be messy and destructive – they like to jump and climb and can be mischievous at times.

Q: Where can I buy pygmy goats?

A: You can find pygmy goats at local goat rescues and you can also purchase them from breeders or online ads. Adopting a goat from a rescue may be the least expensive option but buying from a breeder ensures that the goat is of good breeding quality. Check the website of your national goat club for a list of registered breeders.

Q: What should I do before buying a pygmy goat?

A: Before you actually go out and buy a goat you need to prepare its enclosure. Refer to the zoning requirements for your area to determine how much space you need to provide and build an appropriate shelter for the number of goats you plan to keep. You should also stock up on food for your goats and contact a local veterinarian to ensure that he is available to provide vaccinations. After taking these steps you should find a reputable breeder and contact them about buying a goat.

Q: What questions should I ask before I buy?

A: Before purchasing a goat from a breeder it is a good idea to ask a few questions to ensure that the breeder is knowledgeable and the goat of good breeding quality. Inquire about the age of the goat as well as its health status and history. You may also want to ask what kind of food it is being fed so you can offer the same food during the transition after bringing the goat home.

Breeding Pygmy Goats

Q: Will it be hard to breed my pygmy goats?

A: Pygmy goats breed all year round. In fact, they are very prolific breeders so there is little you have to do to encourage them to breed. It is important to keep males and females separated unless the males are neutered or you intend to use your goats for breeding. Though pygmy goats are capable of breeding at an early age, it is best to wait until they are fully developed.

Q: At what age can I breed my pygmy goats?

A: Pygmy goats can be bred as early as 9 months but it is best to wait until they are 10 to 12 months old. Breeding your pygmy goats too soon could stunt the growth of the fetus or the fetus could actually get stuck during birth, causing the death of both the doe and the fetus.

Q: How many kids can I expect my female goats to produce on a yearly basis?

A: Pygmy goat does typically produce 1 to 4 kids every 9 to 12 months following a 5-month gestation period.

Q: How quickly do newborn kids develop?

A: Pygmy goat kids begin to nurse immediately after being born and they are generally able to run and jump in as little as 4 hours. Within one week of being born, kids generally begin to eat some grain and roughage in addition to nursing.

Q: How can I ensure that there is always a supply of milk for my pygmy goat kids?

A: If you plant to breed your pygmy goats regularly you can breed two pygmy does alternately. This will ensure that while one goat is pregnant, the other is lactating and able to provide milk for growing kids.

Q: At what age should I wean my baby pygmy goats?

A: Pygmy goat kids should not be weaned before 8 weeks of age. The ideal age to wean pygmy goats is between 10 and 12 weeks – by that time they should already be eating some solid food. If the kids are left with their mother, they may continue suckling for up to 6 months.

Q: Do my pygmy goats need to be dehorned and neutered?

A: Dehorning pygmy goats is referred to as disbudding and it is largely a matter of preference. If you do intend to disbud your pygmy goats, it is best to do it between 7 and 14 days after birth. Once the buds attach and begin to grow through the skin, disbudding becomes more stressful and dangerous for the goats. If you don't intend to breed your pygmy goats and simply want to keep them as pets, neutering male goats is not a bad idea. Castrated male goats are referred to as wethers – these goats make great pets and they do not cause the problems stud males are known for.

#

Feeding Pygmy Goats

Q: Do I need to feed my pygmy goats if they have a pasture to graze in?

A: Adequate nutrition is very important for the health of your pygmy goats. If your goats are able to receive adequate nutrition from pasture greens, you may not need to supplement their diet. Young goats as well as lactating and pregnant does have increased energy needs, however, and may require supplemental feeding.

Q: Why isn't my pygmy goat drinking its water?

A: Pygmy goats are clean animals by nature and they may not drink the water you give them if it is not fresh.

Q: How do pygmy goats digest their food?

A: Pygmy goats are ruminant animals – they have stomachs with four different compartments. Like cows, goats chew their food twice – after swallowing it, they regurgitate it later and chew it again. Baby pygmy goats are not born with a fully developed stomach – they do not need one to digest milk. It may take between 8 and 10 weeks for the stomach of a newborn kid to fully develop at which point it can properly digest grains and roughage.

Q: What kind of nutrients do my pygmy goats need?

A: Like all animals, pygmy goats require a balance of carbohydrate, fats and protein to stay healthy. Both carbohydrates and fats provide pygmy goats with energy and protein is important for healthy growth. Certain vitamins and minerals including Vitamins A and D as well as calcium, iodine and phosphorus are also important.

Q: How can I supplement my pygmy goat's diet?

A: To make sure your pygmy goat is getting all the nutrients it needs you may want to consider supplementing its diet with leafy greens and corn. Greens and leafy hay provide pygmy goats with a valuable source of Vitamin A, which is essential for skin and organ health. Corn provides Vitamin A as well and Vitamin D can be supplemented by offering sun-cured hay.

Q: What kind of hay do pygmy goats prefer?

A: Pygmy goats tend to favor legume hays like alfalfa and clover hay over grass hay. Pygmy goats also enjoy leafy plants and weeds but will sometimes graze on wild grasses as well.

Q: What minerals are most important in a pygmy goat's diet?

A: Pygmy goats require calcium, phosphorus, iodine and selenium in their diets. Calcium can be found in alfalfa hay while whole grains provide phosphorus. Iodine can be added to your goat's diet by offering iodized salt and selenium may be given in injection form, if necessary. Your vet should give the injection.

Pygmy Goat Health

Q: How do I keep my pygmy goats from getting sick?

A: Keeping your goats healthy starts with bringing home healthy goats in the first place. Make sure you purchase your goats from a reputable breeder and make sure that the goats have been properly vaccinated and the herd tested for disease. You may also want to have the goats checked out by a veterinarian after you bring them home and get them caught up on any necessary vaccinations. Other than that, you simply need to provide a healthy diet and a clean environment for your goats.

Q: What are the most common diseases affecting pygmy goats?

A: Some of the most common diseases affecting these goats include enterotoxaemia, tetanus, rabies, Johne's disease, sore mouth and parasites. Vaccinations are available and recommended for most of these diseases because they can be very contagious and treatment is not always effective.

Q: How do vaccines work?

A: A vaccine is designed to stimulate your goat's immune system so it is prepared to fight off the particular disease if it is exposed to it at a later stage. The vaccine introduces a very small amount of the disease into your goat's body so its immune system can produce antibodies to attack it. It takes 3 to 4 weeks for your goat to produce sufficient antibodies to protect against the disease and boostering may be required annually.

Q: When should I start vaccinating my pygmy goats?

A: Pygmy goat kids are not born with fully developed immune systems – they receive antibodies from their mother's milk. It takes 8 to 10 weeks for the immune system to develop so it is best to wait until the kid reaches 10 weeks of age to begin vaccination.

Chapter Ten: Relevant Websites

Keeping pygmy goats isn't necessarily difficult, but it is still wise to do your research before you take the plunge. The following websites will help you to find all the information you need to prepare for your pygmy goats and to care for them properly after you bring them home.

In this section you will find websites in the following categories:

Feeding Pygmy Goats
Caring for Pygmy Goats
Health of Pygmy Goats
General Info for Pygmy Goats
Showing Pygmy Goats

1.) Feeding Pygmy Goats

United States Websites:

Bogart, Ralph. "Feeding Pygmy Goats." Agricultural Research Service – United States Department of Agriculture. <http://www.goatworld.com/articles/nutrition/feedingpygmys.shtml>

Kinne, Maxine. "Of Mangers and Feed Plans." Kinne.net. <http://kinne.net/feeding>

White, Jamie. "Feeding and Housing Pygmy Goats." Amber Waves Pygmy Goats. <http://www.amberwavespygmygoats.com/index.php?option=com_content&view=article&id=1179:feeding>

United Kingdom Websites:

"Growing Crops for Goats." PygmyGoat.co.uk. <http://www.pygmygoat.co.uk/Crops.htm>

"Feeding Goats." British Goat Society. <http://www.allgoats.com/feeding>

"Goat Feeding Guide." Small Holder Range. <http://www.smallholderfeed.co.uk/Healthcare-and-Management/Goat-Feeding-Guide-Original.aspx>

2.) Caring for Pygmy Goats

United States Websites:

"How to Care for Your Pygmy's: Basic Condensed Version." TJ's Farms. < http://www.tjsfarms.com/docs/Care.htm>

Hale, Lydia. "Housing." National Pygmy Goat Association. <http://www.npga-pygmy.com/resources/husbandry/housing>

"Pygmy Goat Care." Pegasus Valley. <http://www.pegasusvalley.net/PygmyGoatCare.html>

United Kingdom Websites:

"Goats: Introduction to Welfare and Ownership." RSPCA.org.uk.
<http://www.rspca.org.uk/ImageLocator/LocateAsset?asset=docu
ment&assetId=1232713000349&mode=prd>

"Pygmy Goat – Raising and Keeping Goats." Goodbye City Life.
<http://www.goodbyecitylife.com/animals/pygmy-goat.htm>

"Basic Care Requirements." Pygmy Goat Club.
<http://www.pygmygoatclub.org/general_info/basic_care>

3.) Health of Pygmy Goats

United States Websites:

Getzendanner, Laurie. "Goat Vaccinations." National Pygmy
Goat Association. < http://www.npga-
pygmy.com/resources/health>

"Goat First Aid." Amber Waves Pygmy Goats.
<http://www.amberwavespygmygoats.com/index.php?option=co
m_content&view=article&id=340:goat-health>

Boldrick, Lorrie. "Goat Rumen Illnesses."
UrbanFarmOnline.com.
<http://www.urbanfarmonline.com/urban-livestock>

United Kingdom Websites:

"An Organic Approach to Pygmy Goat Health." Pygmy Goat Club.
<http://www.pygmygoatclub.org/general_info/articles/organic_approach.htm>

"Goat Health." British Goat Society.
<http://www.allgoats.com/health>

Harwood, David. "Goat Health 4 – Responsible Goat Keeping." Nadis.org.uk. <http://www.nadis.org.uk/bulletins/goat-health>

4.) General Info about Pygmy Goats

United States Websites:

"Pygmy Goat." The Oregon Zoo Foundation.
<http://www.oregonzoo.org/discover/animals/pygmy-goat>

"Pygmy Goat." Oklahoma State University Department of Animal Science. <http://www.ansi.okstate.edu/breeds/goats/pygmy/>

"The Pygmy." National Pygmy Goat Association.
<http://www.npga-pygmy.com/resources/husbandry/about_thePygmy.asp>

United Kingdom Websites:

"Pygmy Goats." Goats.co.uk.
<http://www.goats.co.uk/Pygmy_Goats.htm>

"Goat Keeping Information." PygmyGoat.co.uk.
<http://www.pygmygoat.co.uk/Goatinformation.htm>
"Brucklay Pygmy Goats – Pygmy Goat Information."
BrucklayPygmyGoats.co.uk.
<http://www.brucklaypygmygoats.co.uk/Brucklay_Pygmy_Goats_Pygmy_Information.html>

5.) Showing Pygmy Goats

United States Websites:

"Rules for Official Shows." National Pygmy Goat Association.
<http://www.npga-pygmy.com/services/ShowRules.pdf>

Wall, Sandi. "Show Ring Etiquette." The Judging Connection.
http://www.thejudgingconnection.com

"Judging Scorecard for Pygmy Goat Does & Bucks." National
Pygmy Goat Association. <http://www.npga-pygmy.com/resources/conformation>
United Kingdom Websites:

"A Few Hints on Showing Your Pygmy Goat." Pygmy Goat Club.
<http://www.pygmygoatclub.org/general_info/articles/a_few_hints_on_showing.htm>

Butler, Paul. "Showing." Pygmy Goats Ireland.
<http://www.pygmygoatsireland.com/showing.html>

"The Pygmy Goat Breed Standard." Pygmy Goat Club.
<http://www.pygmygoatclub.org/forms/Microsoft%20Word%20-
%20THE%20PYGMY%20GOAT%20BREED%20STANDARD
2012

Index

References

"A Beginner Guide to Kidding." Pygmy Goat Club.
<http://www.pygmygoatclub.org/general_info/kiddingarticle.htm

Blackburn, Lorrie. "Normal Values." National Pygmy Goat
Association. <http://www.npga-
pygmy.com/resources/health/normal_values.asp>

Blankevoort, Mary. "Parasitism in Pygmy Goats." National
Pygmy Goat Association. <http://www.npga-
pygmy.com/resources/health/parasitism.asp>

Blankevoort, Mary. "Pneumonia in Goats." National Pygmy Goat
Association. <http://www.npga-
pygmy.com/resources/health/pneumonia.asp>

Bogart, Ralph. "Feeding Pygmy Goats." Agricultural Research
Service – United States Department of Agriculture.
<http://www.goatworld.com/articles/nutrition/feedingpygmys.sht
ml>

"Cost of Raising a Goat." Irvine Mesa Charros 4-H Club.
<http://www.goats4h.com/Goat-costs>

"Dermatitis." Goat World.
<http://www.goatworld.com/articles/dermatitis/>

"Edible and Poisonous Plants for Goats." Fias Co Farm.
<http://fiascofarm.com/goats/poisonousplants.htm>

Everett, Nic. "Guidelines on Pygmy Goat Color Requirements and Descriptions for NPGA Registration." National Pygmy Goat Association. <http://www.npga-pygmy.com/resources/conformation

"Feeding Goats." Fiasco Farm. <http://fiascofarm.com/goats/feeding.htm>

Gasparotto, Suzanne. "Getting Goat Nutrition Right." Onion Creek Ranch. <http://www.tennesseemeatgoats.com/articles2/feedinggoatsproperly.html>

Getzendanner, Laurie. "Goat Vaccinations." National Pygmy Goat Association. < http://www.npga-pygmy.com/resources/health>

"Guidance for Local Authorities on the Licensing of Movements of Livestock." Gov.UK. <https://www.gov.uk/government/publications/guidance-for-local-authorities-on-the-licensing-of-movements-of-livestock>

Hale, Lydia. "Housing." National Pygmy Goat Association. <http://www.npga-pygmy.com/resources/husbandry/housing>

"Holding Register." Gov.UK. <https://www.gov.uk/government/uploads/system/uploads/attachment_data/file/69416/pb13281-holding-register-091209.pdf>

"How to Care for Your Pygmys: Basic Condensed Version." TJ's Farms. < http://www.tjsfarms.com/docs/Care.htm>

"Judging Scorecard for Pygmy Goat Does and Bucks." National Pygmy Goat Association. < http://www.npga-pygmy.com/resources/conformation>

"Judging Scorecard for Showmanship." National Pygmy Goat Association. < http://www.npga-pygmy.com/resources/conformation

Kinne, Maxine K. "Pygmies for All Reasons." National Pygmy Goat Association. < http://www.npga-pygmy.com/resources/husbandry/allreasons.asp>

Kinne, Maxine. "Smoke Gets in Your Eyes." Kinne.net. <http://kinne.net/disbud.htm>

Kinne, Maxine. "Poisonous Plants and Toxic Substances." National Pygmy Goat Association. < http://www.npga-pygmy.com/resources/health/poisonous_plants.asp>

Krieg, Elaine. "Information on the NPGA Johne's Health Alert." National Pygmy Goat Association. < http://www.npga-pygmy.com/resources/health>

Leman, Maggie. "For Your Information: Pygmy Goats 101." Keystone Pygmy Goat Club. <http://www.keystonepygmygoatclub.com/fyi.htm>

Leman, Maggie. "How to Buy a Pygmy Goat: Questions Buyers Should Ask." National Pygmy Goat Association. <http://www.npga-pygmy.com/resources/husbandry/buying_goat.asp>

Lewis, Mary Ann. "Playground for Pygmies." National Pygmy
Goat Association. < http://www.npga-
pygmy.com/resources/husbandry/playground>

"Livestock." Waterville Municipal Code.
<http://www.codepublishing.com/wa/waterville/html/waterville06
/waterville0620.html>

Maas, Jennifer. "Urinary Calculi." National Pygmy Goat
Association. <http://www.npga-
pygmy.com/resources/health/urinary_calculi.asp>

"Miniature, Dwarf or Pygmy Goats." Farm Alliance
Baltimore.org. < http://www.farmalliancebaltimore.org/goats/>

"NPGA Breed Standard." National Pygmy Goat Association.
<http://www.npga-pygmy.com/resources/conformation>

Orlando, Kay. "First Aid." National Pygmy Goat Association.
<http://www.npga-pygmy.com/resources/health/firstaid.asp>

Orlando, Kay. "Vitamin and Mineral Supplements." National
Pygmy Goat Association. < http://www.npga-
pygmy.com/resources/health/vitamins>

Panhwar, Farzana. "The Common Diseases of Goats." Goat
World.
<http://www.goatworld.com/articles/health/commondiseases.shtm
l>

Pavia, Audrey. "Keeping Goats as Companions."
UrbanFarmOnline.com.
<http://www.urbanfarmonline.com/urban-livestock>

"Pygmy Goat." BioExpedition.com.
<http://bioexpedition.com/pygmy-goat/>

"Pygmy Goat." Oklahoma State University Department of Animal
Science. <http://www.ansi.okstate.edu/breeds/goats/pygmy/>

"Pygmy Goat." The Oregon Zoo Foundation.
<http://www.oregonzoo.org/discover/animals/pygmy-goat>

"Pygmy Goats Make Excellent and Unique Pets." Amber Waves
Pygmy Goats.
<http://www.amberwavespygmygoats.com/index.php?option=co
m_content&view=article&id=1775:pygmy-goats-make-excellent-
and-unique-pets-&catid=902:first-time-goat-buyer&Itemid=73>

"Rabies." Health Central.
<http://www.healthcentral.com/encyclopedia/408/738.html>

Schoenian, Susan. "General Health Care of Sheep and Goats."
Amber Waves Pygmy Goats."
<http://www.amberwavespygmygoats.com/index.php?option=co
m_content&view=article&id=1930%3Ageneral-health>

Schoenian, Susan. "Soremouth (ORF) in Sheep and Goats." Small
Ruminant Info Sheet – University of Maryland Extension.
<http://www.sheepandgoat.com/articles/soremouth.html>

Schoenian, Susan. "Vaccinations for Sheep and Goat Flocks."
Amber Waves Pygmy Goats.
<http://www.amberwavespygmygoats.com/index.php?option=co
m_content&view=article&id=1019%3Avaccinations-for-sheep-
and-goat-flocks&catid=21%3Adiseases&Itemid=73>

"Sheep and Goat Keepers – England Important Information."
Gov.UK.
<https://www.gov.uk/government/uploads/system/uploads/attachment_data/file/69430/pb13441-sheep-goat-individual-reporting-reqd-101110.pdf>

Talley, Justin. "External Parasites of Goats." Oklahoma State
University.
<http://pods.dasnr.okstate.edu/docushare/dsweb/Get/Document-5175/EPP-7019web.pdf>

"Tetanus." Onion Creek Ranch.
<http://www.tennesseemeatgoats.com/articles2/tetanus.html>

"The Biology of the Goat." Goat Biology.
<http://www.goatbiology.com/parasites.html>

"The Pygmy." National Pygmy Goat Association.
<http://www.npga-pygmy.com/resources/husbandry/about_thePygmy.asp>

"The Pygmy Goat Breed Standard." Pygmy Goat Club.
<http://www.pygmygoatclub.org/forms/Microsoft%20Word%20-%20THE%20PYGMY%20GOAT%20BREED%20STANDARD2012.pdf>

"The Pygmy Goat Club: How it Began." Pygmy Goat Club.
<http://www.pygmygoatclub.org/general_info/articles/pgc_howitbegan.htm>

Thompson, Margaret. "The Male Goat." Pygmy Goat Club.
<http://www.pygmygoatclub.org/general_info/themalegoat.htm>

Van Metre, D. "Enterotoxemia (Overeating Disease) of Sheep and Goats." Colorado State University Extension. <http://www.ext.colostate.edu/pubs/livestk/08018.html>

Van Saun, Robert. "Parasites in Goats." Department of Veterinary Science Penn State University. <http://vbs.psu.edu/extension/resources/pdf/presentations/PR-Goat-internal-parasites-VanSaun.pdf>

Walters, N. Galen. "Coccidiosis: Understanding the Drugs Available for Control." National Pygmy Goat Association. <http://www.npga-pygmy.com